Nicolás Arturo Osorio Gracia
Carlos Nick Gomes
Edgard A . Picoli

Production and quality of mechanically pollinated tomatoes

Nicolás Arturo Osorio Gracia
Carlos Nick Gomes
Edgard A . Picoli

Production and quality of mechanically pollinated tomatoes

Natural and mechanical pollination in two planting seasons and three tomato varietal groups

ScienciaScripts

Imprint

Any brand names and product names mentioned in this book are subject to trademark, brand or patent protection and are trademarks or registered trademarks of their respective holders. The use of brand names, product names, common names, trade names, product descriptions etc. even without a particular marking in this work is in no way to be construed to mean that such names may be regarded as unrestricted in respect of trademark and brand protection legislation and could thus be used by anyone.

Cover image: www.ingimage.com

This book is a translation from the original published under ISBN 978-613-9-70216-9.

Publisher:
Sciencia Scripts
is a trademark of
Dodo Books Indian Ocean Ltd. and OmniScriptum S.R.L publishing group

120 High Road, East Finchley, London, N2 9ED, United Kingdom
Str. Armeneasca 28/1, office 1, Chisinau MD-2012, Republic of Moldova, Europe
Printed at: see last page
ISBN: 978-620-8-17530-6

ACKNOWLEDGEMENTS

To the Federal University of Viçosa and the Postgraduate Programme in Plant Science, for the opportunity to study for a master's degree.

To Professor Carlos Nick Gomes, for his guidance and friendship.

To Professor Edgard Augusto de Toledo Picoli and Derly José Henriques da Silva, for their guidance and valuable suggestions.

To my friends at the Olericulture Studies Centre (NEO) for their friendship and help over the last few years.

To my parents, Acened Gracia Angarita and Arturo Osorio Arevalo, for all their financial support, strength, trust and unconditional love and affection.

To all my family for their support.

And to everyone else who contributed in some way to the completion of this work.

Thank you very much!

Nicolás Arturo Osorio Gracia, son of Acened Gracia Angarita and Arturo Osorio Arevalo, was born on 15 October 1991 in the city of Ibagué, department of Tolima, Colombia.

In February 2009, he began his Agronomy degree at the Universidad del Tolina in Ibagué, Colombia, graduating in March 2014.

In August 2014, he joined the Postgraduate Programme in Plant Science, at Master's level, at the Federal University of Viçosa, where he defended his dissertation in August 2016.

SUMMARY

SUMMARY

OSORIO GRACIA, Nicolás Arturo, M. Sc., Federal University of Viçosa, August 2016. **Production and quality of tomatoes from natural and mechanical pollination.** Supervisor: Carlos Nick Gomes. Co-supervisors: Derly José Henriques da Silva and Edgard Augusto de Toledo Picoli.

The tomato plant is recognised as a model plant for studying the development of the fruit of other species that have anthers with poricidal dehiscence, a characteristic that hinders the pollination process. Mechanical pollination is a practice that facilitates the release of pollen, improving fruit production and quality. However, little information about the efficiency of this type of pollination in fruit production restricts its use. With this in mind, the aim of this proposal is to study the effects of mechanical pollination on the production and quality of tomatoes from three commercial varietal groups intended for fresh consumption in two planting seasons. The experiment was conducted in the field at the Horticultural Research and Extension Unit of the Phytotechnics Department at the Federal University of Viçosa. The experimental design was randomised blocks, in a 2 (planting times) x 2 (natural pollination and mechanical pollination) factorial scheme. The trials were carried out in autumn/winter and summer/autumn for each group, evaluating them separately. Variables relating to yield, quality, dry mass of seeds and number of seeds were assessed. There was an increase in productivity with the use of mechanical pollination in both planting seasons for the Salada and Italiano varietal groups, while the Santa Cruz group only saw this increase in the autumn/winter season. Mechanical pollination only increased the number of seeds in the Salada group. Regardless of the planting season for the Salada and Italiano varietal groups, mechanical pollination works efficiently to increase production without altering tomato quality.

CHAPTER 1

As well as being an important source of nutrients in the human diet due to its functional properties, the tomato plant is recognised as a model plant for studying the development of fleshy fruits due to its genetic characteristics, such as a short phenological cycle (TOHGE; FERNIE, 2015) and wide climatic adaptation (DE CÁSSIA PEREIRA-CARVALHO et al., 2014).

Pollination is a symbiotic process that occurs between pollinating and pollinated species, and its importance lies in ensuring reproduction, fruit development and maintaining the genetic diversity of a wide variety of foods, especially vegetable crops (FAO, 2015).

Techniques such as mechanical pollination have shown positive results in improving the yield and quality of fruit in the tomato production system. Despite the importance of the crop, little has been explored in terms of the theoretical basis that justifies the increase in attributes related to quality and yield resulting from physiological and morphological processes in the fruit. This information is necessary for a critical and coherent understanding and interpretation of the tomato fruit production process in order to apply and maximise its use on a production scale.

Most solanaceous plants have poricidal dehiscence, which makes the pollination process a determining factor in the development of the fruit, since the tomato flowers need to be vibrated by the pollinating agents to facilitate the release of pollen (CARRIZO GARCÍA et al., 2008).

Fruit development is determined by morphological and physiological interactions through the action of hormones responsible for activating enzymes and proteins that favour the growth, elongation, expansion and multiplication of cells that make up the fruit's tissues. These interactions are also related to the chemical characteristics that confer flavour, aroma and colour, and which affect the production and final quality of the tomato (WHITE, 2002).

Current studies aimed at quantifying the quality and productivity of tomato plants relate these characteristics to the crop's soil and climate conditions (MUELLER et al., 2013; KITTAS et al., 2015), plant architecture (HEINE et al., 2015; TAKAHASHI; CARDOSO, 2015; VIVIAN et al., 2015), post-harvest activities (PEREIRA et al., 2013), propagation methods (TEIXEIRA, 2014; JAVANMARDI; MORADIANI, 2016) and pollination by bees in protected environments (PRESSMAN et al., 1999; SILVA et al., 2010; DE; VALLEJO-MARIN, 2013; SANTOS et al., 2014). However, few studies have looked at the influence of mechanical pollination on crops grown in full sun, as a complementary practice that determines the quality and yield of tomato production, explained in morphological and physiological terms during each stage of fruit development and under different environmental conditions.

The importance of studying the effects of mechanical pollination on fruit development implies obtaining preliminary information when applying phytotechnical knowledge, using the stages of development after anthesis as a standard or point of reference. The aim of this proposal is to evaluate the effect of mechanical pollination on the production and quality of tomato fruit from three commercial hybrids grown in the field at different planting times.

2. LITERATURE REVIEW

2.1. Tomato cultivation

The tomato *(Solanum lycopersicum* L.) is a species of the solanaceae family, native to the western region of South America; it occupies a wide range of environmental conditions such as arid, humid and mountainous regions, at altitudes of more than 3,300 metres above sea level (DE CÁSSIA PEREIRA-CARVALHO et al., 2014).

It is an autogamous plant with regular, hypogynous flowers, with the number of flowers varying from 6 to 15 depending on the variety, hermaphrodite, in the shape of a raceme. The fruits are fleshy, juicy berries and can be bi- or plurilocular (depending on the composition of the ovary) (FONTES; SILVA, 2002). The fruit is climacteric and develops from a 5 to 10 mg ovary, reaching a final mass of 5 to 500 grams when ripe, depending on the variety (ALVARENGA, 2013).

The main tomato groups on the Brazilian market are Salada, Italiano, Santa Cruz and Cereja, with a share of 52.2%, 25.1%, 21.9% and 0.8% respectively (PEREIRA-CARVALHO et al., 2014). The crop plays an important role in the national economy, with an annual turnover of approximately R$4.2 billion and generating more than 650,000 jobs in production systems (MELO, 2014).

2.2. Fruit development

Fruit formation is directly related to the number of seeds, the number of locules (ALVARENGA, 2013), the number and volume of cell layers in the pericarp, which depend on the degree of cell division and expansion (ARIIZUMI et al., 2013).

Tomato development begins when the ovule is fertilised and requires coordination between the development of the seeds and the growth and differentiation of the ovary. Fertilisation produces drastic changes in the phytohormonal balance, which through synchronised events activate the processes and mechanisms that lead the fruit to develop (DORCEY et al., 2009; OBROUCHEVA, 2014).

Based on histochemical and physiological considerations, GILLASPY et al. (1993) proposed that tomatoes develop in four phases over a period of 40 to 70 days.

The first phase is characterised by intense mitotic activity and begins with flower development where the carpel is completely formed and ends when the flower reaches anthesis, during which time the oospheres are fertilised. The second phase is the slowest, the fruit reaches only 10 per cent of its final fresh weight, undergoing multiple cell divisions during 7 to 14 days after anthesis. The third phase is marked by cell expansion, which takes 3 to 5 weeks before the fruit reaches its normal size. In the fourth phase, the fruit undergoes physical and chemical changes, giving it its characteristic texture, colour and flavour.

Temperature is one of the environmental factors that has the greatest effect on fruit development, affecting the processes of cell division and expansion corresponding to phases two and three, respectively. Fruits that develop at high temperatures have higher growth rates compared to fruits that develop at low temperatures (BERTIN, 2005; OKELLO et al., 2015), According to BERTIN (2005), fruit developed at 25 °C had a lower number of cells but a larger size as a result of a shorter cell division phase, compared to fruit that developed at 20 °C, which had a higher number of cells with a reduced size due to the long cell division period; in this study, the temperature did not affect the size of the fruit due to a compensation between the number and size of the cells in the cell division and expansion phases respectively.

Fruit development can be explained from its anatomy in the different phases, giving a broad holistic approach to the hormonal and physiological events that take place. Plant growth curves can be a way of elucidating these events, and their behaviour adjusts according to the conditions of the ecosystem (POSADA et al., 2007).

2.3. Pollination

Crop yield depends on morphophysiological and biochemical processes depending on the environment (RANCIC et al., 2010). Pollination is a determining environmental service for fruit production in most angiosperms, as the start of fruiting depends on the joint success of this process and the fertilisation of the oospheres (ARIIZUMI et al., 2013). A poor pollination process leads to losses in production, with the formation of small and defective fruit (HIGUTI et al., 2010). According to MORANDIN et al. (2001), the amount of pollen transferred to the stigma determines the size of the tomato, a consequence of greater fertilisation of the oospheres and the formation of seeds that will promote the production of hormones involved in the processes of fruit division, expansion and ripening (SHINOZAKI et al., 2015).

The dehiscence of the anthers is an essential step that determines the pollination of the tomato plant, where the pollen grains are released after ripening (CARRIZO GARCÍA et al., 2008). Tomato flowers have anthers with poricidal dehiscence, which is a limiting characteristic for pollen release, as the anther opens at the top at a single point (CARRIZO GARCÍA et al., 2008; SILVA et al., 2010; DE; VALLEJO-MARIN, 2013). According to SILVA et al. (2010), tomato flowers need to be vibrated to facilitate the release of pollen through the anther pores and subsequent deposition on the stigma so that the oosphere can be fertilised.

In tomato plants, the main pollinating agent are bees of the genus *Bombus* (VELTHUIS, 2002; VERGARA; FONSECA-BUENDÍA, 2012), which induce the release of pollen through vibrations (*buzz pollination*) made by the insect's thoracic movements when they are on the anthers of the flowers (DE; VALLEJO-MARIN, 2013; GOULSON, 2010; SILVA et al., 2010). However, a complementary alternative to bees would be to use mechanical mechanisms that can replace the function performed by natural pollinators when environmental conditions are unfavourable (GOULSON, 2010). Among the main unfavourable environmental conditions are low temperatures and high relative humidity (PRESSMAN et al., 1999).

2.4. Mechanical pollination

Among the main types of mechanical pollination used on tomato plants are: mechanical blowers by tractor or backpack (NAHIR et al., 1984), manual balancing of trellis ties, individual balancing of plants (HIGUTI et al., 2010) and electrical vibration of flowers (CUÉLLAR et al., 2011; PRESSMAN et al., 2015).

The good performance of mechanical pollination is based on the efficiency of vibratory frequencies, where this method can release a greater quantity of pollen compared to natural agents. HARDER; BARCLAY (1994) reported that mechanical pollination manages to release 23.4 % of the pollen in a single vibration with frequencies between 400 and 1,000 Hz, while in a single visit the Bombus bees only released 9.4 % of the pollen with frequencies below 400 Hz. According to PRESSMAN et al. (1999) mechanical pollination can be equal to or superior to pollinating insects when environmental conditions are unfavourable to the action of these insects.

Mechanical pollination allows farmers to increase productivity and ensure greater profit when their crops are at risk due to phytosanitary problems, adverse weather conditions or low market prices. This practice is advantageous because its establishment does not imply any changes to the usual cultural practices such as spacing, planting, irrigation systems, among other crop maintenance activities, and it is possible to obtain benefits immediately after its implementation. Furthermore, the use of mechanical pollination methods does not result in negative impacts on the agro-ecosystem and favours the sustainability of tomato plants.

The majority of studies on mechanical pollination systems for tomato plants are carried out under protected growing conditions, since the lack of air currents and the low incidence of pollinating insects in these places provide an environment that is not very favourable to flower pollination (HARDER; BARCLAY, 1994; ALDANA et al., 2007; HARDER; HIGUTI et al., 2010; CUÉLLAR et al., 2011; SANTOS, 2014). In this context, it would be necessary to carry out studies that verify the behaviour of mechanical pollination in field conditions, in order to elucidate the effects and agronomic contributions to increasing productivity.

3.   REFERENCES

ALDANA, J.; CURE, J. R.; ALMANZA, M. T.; VECIL, D.; RODRÍGUEZ, D. Efecto de *Bombus atratus* (Hymenoptera: Apidae) sobre la productividad de tomate

(Lycopersicon esculentum Mill.) bajo invernadero en la Sabana de Bogotá, Colombia. **Agronomia colombiana**, v. 25, n. 1, p. 62-72, 2007.

ALVARENGA, M. A. R. Tomato: production in the field, greenhouse and hydroponics. Rev2. **Lavras: Editora Universitária de Lavras**, p. 19, 2013.

ARIIZUMI, T.; SHINOZAKI, Y.; EZURA, H. Genes that influence yield in tomato. **Breeding Science**, v. 63, n. 1, p. 3-13, 2013.

BERTIN, N. Analysis of the tomato fruit growth response to temperature and plant fruit load in relation to cell division, cell expansion and DNA endoreduplication. **Annals of Botany**, v. 95, n. 3, p. 439-447, 2005.

CARRIZO GARCIA, C.; MATESEVACH, M.; E BARBOZA, G. Features related to anther opening in *Solanum* species (Solanaceae). **Botanical Journal of the Linnean Society**, v. 158, n. 2, p. 344-354, 2008.

CUÉLLAR, J; COOMAN, A; ARJONA, H. Increasing the yield of overwintered tomato crops by improving pollination. **Agronomía Colombiana**, v.

18, n. 1-3, p. 39-45, 2001.

DE CÁSSIA PEREIRA-CARVALHO, R.; TOBAR, L. L. M.; DE CAMPOS DIANESE, E.; DE NORONHA FONSECA, M. E.; BOITEUX, L. S.L. Tomato genetic improvement for resistance to diseases of viral etiology: advances and perspectives. **RAPP**, v. 22, p. 280-361, 2014.

DE LUCA, P. A.; VALLEJO-MARIN, M. What's the 'buzz'about? The ecology and evolutionary significance of buzz-pollination. **Current Opinion in Plant Biology**, v. 16, n.

4, p. 429-435, 2013.

DORCEY, E.; URBEZ, C.; BLÁZQUEZ, M. A.; CARBONELL, J.; PEREZ-AMADOR, M. A. Fertilisation-dependent auxin response in ovules triggers fruit development through the modulation of gibberellin metabolism in Arabidopsis. **The Plant Journal**, v. 58, n. 2, p. 318-332, 2009.

FAO. **Biodiversity group**: Pollinators. 2015. Available at: <http://www.fao.org/biodiversity/components/pollinators/en/>. Accessed in July 2016.

FONTES, P.; SILVA, D. D. Diseases and pests: is it safe to eat tomatoes? **Table Tomato Production. Viçosa: Aprenda Fácil**, p. 97-129, 2002.

GILLASPY, G.; BEN-DAVID, H.; GRUISSEM, W. Fruits: a developmental perspective. **The Plant Cell**, v. 5, n. 10, p. 1439, 1993.

GOULSON, D. Bumblebees: behaviour, ecology and conservation 2 ed. **Oxford University Press**, 2010.

HARDER, L. D.; BARCLAY, R. M. R. The functional significance of poricidal anthers and buzz pollination: controlled pollen removal from Dodecatheon. **Functional Ecology**, p. 509-517, 1994.

HEINE, A. J. M.; MORAES, M. O. B.; PORTO, J. S.; SOUZA, J. R.; REBOUÇAS, T. N. H.; SANTOS, B. S. R. Stem number and spacing in tomato production and quality. **Scientia Plena**, v. 11, n. 9, 2015.

HIGUTI, A. R. O.; GODOY, A. R.; SALATA, A. D. C.; CARDOSO, A. I. I. Tomato production in function of plant" vibration". **Bragantia,** v. 69, n. 1, p. 87-92, 2010.

JAVANMARDI, J.; MORADIANI, M. Tomato Transplant Production Method Affects Plant Development and Field Performance. **International Journal of Vegetable Science,** p. 1-11, 2016.

KITTAS C.; KATSOULAS N.; RIGAKIS V.; BARTZANAS T.; KITTA E. Effects on microclimate, crop production and quality of a tomato crop grown under shade nets. **The Journal of Horticultural Science and Biotechnology,** v. 87, n. 1, p. 7-12, 2012.

MELO, PAULO CT. Recent advances in table tomato growing associated with changes in the technological paradigm and challenges to overcome. **5th National Table Tomato Seminar (5th SNTM).** Available at <http://www.tomatedemesa.com.br/2014/noticia01.html>. UNIMEP, Piracicaba-SP, 2014.

MUELLER, S.; WAMSER, A. F.; SUZUKI, A.; BECKER, W. F. Tomato productivity under organic fertilisation and supplementation with mineral fertilisers. **Horticultura Brasileira,** v. 31, n. 1, p. 86-92, 2013.

MORANDIN, L. A.; LAVERTY, T. M.; KEVAN, P. G. Effect of bumble bee (Hymenoptera: Apidae) pollination intensity on the quality of greenhouse tomatoes. **Journal of Economic Entomology,** v. 94, n. 1, p. 172-179, 2001.

NAHIR, D.; GAN-MOR, S.; RYLSKI, I.; FRANKEL, H. Pollination of Tomato Flowers by a Pulsating Air Jet. **Transactions of the ASAE,** v. 27, n. 3, p. 894-896, 1984.

OBROUCHEVA, N. V. Hormonal regulation during plant fruit development. **Russian Journal of Developmental Biology,** v. 45, n. 1, p. 11-21, 2014.

OKELLO, R. C.; DE VISSER, P. H.; HEUVELINK, E.; LAMMERS, M.; DE MAAGD, R. A.; STRUIK, P. C.; MARCELIS, L. F. A multilevel analysis of fruit growth of two tomato cultivars in response to fruit temperature. **Physiologia Plantarum**, v. 153, n. 3, p. 403-418, 2015.

PEREIRA RAMOS, A. R.; ESTEVES AMARO, A. C.; MACEDO, A. C.; ASSIS SUGAWARA, G. S. D.; EVANGELISTA, R. M.; RODRIGUES, J. D.; ONO, E. O. Quality of 'giuliana' tomato fruit treated with products with physiological effects. **Seminar: Agricultural Sciences**, p. 3543-3552, 2013.

POSADA, F. C.; CARDOZO, M. C.; HERNÁNDEZ, J. C. Análisis del crecimiento en frutos de tomate (Lycopersicon esculentum Mill.) cultivados bajo invernadero. **Agronomía colombiana**, v. 25, n. 2, p. 299-305, 2007.

PRESSMAN, E.; SHAKED, R.; ROSENFELD, K.; HEFETZ, A. A comparative study of the efficiency of bumble bees and an electric bee in pollinating unheated greenhouse tomatoes. **The Journal of Horticultural Science and Biotechnology**, v. 74, n. 1, p. 101-104, 1999.

RANCIC, D.; QUARRIE, S. P.; PECINAR, I. Anatomy of tomato fruit and fruit pedicel during fruit development. **Microscopy: Science, Technology, Applications and Education, Méndez-Vilas A, Díaz J (Eds), Formatex, Badajoz, Spain**, p. 851861, 2010.

SANTOS, A. O. R.; BARTELLI, B. F.; NOGUEIRA-FERREIRA, F. H. Potential pollinators of tomato, Lycopersicon esculentum (Solanaceae), in open crops and the effect of a solitary bee in fruit set and quality. **Journal of Economic Entomology**, v. 107, n. 3, p. 987-994, 2014.

SHINOZAKI, Y.; HAO, S.; KOJIMA, M.; SAKAKIBARA, H.; OZEKI-IIDA, Y.;

ZHENG, Y.; OKABE, Y. Ethylene suppresses tomato (Solanum lycopersicum) fruit set through modification of gibberellin metabolism. **The Plant Journal**, v. 83, n. 2, p. 237-251, 2015.

SILVA, P. N.; HRNCIR, M.; FONSECA, V. L. I. Pollination by vibration. **Oecologia Australis**, v. 14, n. 1, p. 140-151, 2010.

TAKAHASHI, K.; CARDOSO, A. I. Production and quality of mini tomatoes in an organic system with two types of stem guidance and apical pruning. **Horticultura Brasileira**, v. 33, n. 4, p. 515-520, 2015.

TEIXEIRA, J. M. D. S. 2014, 61f. Evaluation of the system of tomato (Lycopersicum esculentum Mill) grafting in protected culture on the productivity and quality of the fruit. Master's dissertation in Organic Agriculture - **Escola Superior Agrária do Instituto Politécnico de Viana do Castelo**. 2014.

TOHGE, T.; FERNIE, A. R. Metabolomics-Inspired Insight into Developmental, Environmental and Genetic Aspects of Tomato Fruit Chemical Composition and Quality. **Plant & cell physiology**, v. 56, n. 9, p. 1681-96, sep. 2015.

VELTHUIS, H. H. The historical background of the domestication of the bumble-bee, Bombus terrestris, and its introduction in agriculture. Pollinating Bees-The conservation link between agriculture and nature. **Ministry of Environment**, São Paulo, Brazil, p. 177-184, 2002.

VERGARA, C. H.; FONSECA-BUENDÍA, P. Pollination of greenhouse tomatoes by the Mexican bumblebee Bombus ephippiatus (Hymenoptera: Apidae). **Journal of Pollination Ecology**, v.7, p. 27-307, 2012.

VIVIAN, R.; ROCHA, A.; GALVÃO, H. L.; MARTINEZ, H. E. P.; PEREIRA, P. R. G.; FONTES, P. C. R. Planting density and number of leaves influencing yield and fruit quality of tomato plants grown with one cluster in a hydroponic system. **Ceres**, v. 55, n. 6, 584-589, 2015.

WHITE, P. J. Recent advances in fruit development and ripening: an

overview. **Journal of Experimental Botany**, v. 53, n. 377, p. 1995-2000, 2002.

CHAPTER 2

PRODUCTION AND QUALITY OF TOMATOES FROM NATURAL AND MECHANICAL POLLINATION

SUMMARY

Machine pollination is a practice used to increase the productivity and nutritional quality of tomato fruit, but there is still a lack of information to support its agronomic viability in production systems. With this in mind, the aim of this proposal is to study the effects of mechanical pollination on the production and quality of tomatoes from three commercial varietal groups intended for fresh consumption in two planting seasons. The experiment was conducted in the field at the Horticultural Research and Extension Unit of the Phytotechnics Department at the Federal University of Viçosa. The experimental design was randomised blocks, in a 2 (planting times) x 2 (natural pollination and mechanical pollination) factorial scheme. The trials were carried out in autumn/winter and summer/autumn for each group, evaluating them separately. Variables relating to yield, quality, dry mass of seeds and number of seeds were assessed. An increase in yield was observed with the use of mechanical pollination in both planting seasons for the Salada and Italiano varietal groups, while the Santa Cruz group only saw this increase in the autumn/winter season. Mechanical pollination only increased the number of seeds in the Salada group. Regardless of the planting season for the Salada and Italiano varietal groups, mechanical pollination works efficiently to increase production without altering tomato quality.

Keywords: Solanum lycopersicum, productivity and sustainability

3.1. INTRODUCTION

The tomato plant is recognised as a model plant for the study of fleshy fruits (TOHGE; FERNIE, 2015). This species has climacteric fruits that develop from a 5 to 10 mg ovary, reaching a final fresh mass of 5 to 500 g when ripe, depending on the variety (ALVARENGA, 2013).

Among the main tomato variety groups on the Brazilian market are Salada, Italiano, Santa Cruz and Cereja with a share of 52.2%, 25.1%, 21.9% and 0.8% respectively (PEREIRA-CARVALHO et al., 2014). The crop plays an important role in the national economy, with an annual turnover of approximately R$4.2 billion and generating more than 650,000 jobs in production systems (MELO, 2014).

Tomato yield is determined by the efficiency of production and final fruit formation (ARIIZUMI et al., 2013) and is directly related to the number of seeds, the number of locules (ALVARENGA, 2013) and the number and volume of pericarp cell layers, which depend on the degree of cell division and expansion (ARIIZUMI et al., 2013). These events are the result of molecular and physiological processes regulated by hormones (GILLASPY et al., 1993) which act as modulators that synchronise the cell cycle; auxins, gibberellin, cytokinin, abscisic acid and ethylene being the most important (SRIVASTAVA; HANDA, 2005).

Different factors such as cultivar, temperature, light, relative humidity, nutrition and the effects of growth regulators favour flower development and contribute to fruit formation and quality (ALVARENGA, 2013).

On the other hand, pollination is an environmental service that directly influences the development of the fruit, favouring its formation, quality and productivity, since the start of fruiting depends on the joint success of pollination and fertilisation of the ovules (ARIIZUMI et al., 2013; GILLASPY et al., 1993).

There are few studies on the effects of pollination on the physicochemical properties of the fruit. For the flowers of most solanaceous plants, this is a vital process due to the morphology of their anthers when they are close to releasing pollen (CARRIZO GARCÍA et al., 2008; SILVA et al., 2010). In tomatoes, the dehiscence (poricide) of the anthers is a limiting factor in the pollination process, since the opening and arrangement of the stomata (the point where the dehiscence occurs) takes place through apical pores and at a single point, damming up the pollen grains and hindering their release (CARRIZO GARCÍA et al., 2008). According to SILVA et al. (2010), tomato flowers need to be vibrated to facilitate the release of pollen through the anther pores and allow the grains to be deposited on the stigma. CARDOSO (2007) argues that this process under protected growing conditions increases fruit set, diameter and quality.

In tomatoes, the main pollinating agent are bees of the genus *Bombus* (VELTHUIS, 2002; VERGARA; FONSECA-BUENDÍA, 2012), which induce the release of pollen through vibrations (*buzz pollination*) made by the insect's thoracic movements when they are on the anthers of the flowers (DE; VALLEJO-MARIN, 2013; GOULSON, 2010; SILVA et al., 2010).

Low insect vibration frequencies (below 400Hz) (HARDER; BARCLAY, 1994; GOULSON, 2010) and unfavourable environmental conditions related to high relative humidity and low temperatures (PRESSMAN et al., 1999; VANBERGEN; BAUDE, 2013) hinder the natural pollination process. A complementary alternative to the function performed by insects and the wind is to use mechanical vibrators that optimise the function performed by natural pollinators, as these mechanisms have the capacity to carry out greater vibrations (greater than 400 Hz) in a constant and selective manner, facilitating the release and attachment of pollen (HARDER; BARCLAY, 1994; GOULSON, 2010).

There are various anthropogenic methods that can enhance the pollination process, among them: blowers (mechanical or bacpack) (NAHIR et al., 1984), manual balancing of ribbons, balancing of the plant guard (HIGUTI et al., 2010) and vibration by electric toothbrush (CUÉLLAR et al., 2011; PRESSMAN et al., 1999).

There are several studies on the effects of pollination on tomato production in protected environments, but there is limited information on the possible changes in fruit quality and productivity under other conditions, i.e. tomato cultivation in full sun

(ALDANA et al., 2007; HIGUTI et al., 2010; SANTOS, 2014). For this reason, the aim of our proposal is to carry out a comparative study of the production and quality of tomato fruit from natural and mechanical pollination at field level and to understand their effects at different planting times.

4.2   MATERIAL AND METHODS

Trials were carried out in autumn/winter and summer/autumn to check the effects of mechanical pollination on the production and quality of Santa Cruz, Salada and Italiano tomatoes, evaluating them separately. The trials were conducted in the Research Garden of the Federal University of Viçosa, in the municipality of Viçosa, Minas Gerais, in an area of 384 m^2 . During the flowering phase in the autumn/winter crop, the average temperature was 18 °C and the average relative humidity was 83 %, while in the summer/autumn it was 23 °C and 78 % respectively.

The experimental design for the trials was randomised blocks, in a 2 x 2 factorial scheme, with four replications. The treatments were made up of a combination of two planting seasons: autumn/winter, summer/autumn; and two types of pollination: natural and mechanical. The experimental plots consisted of ten plants, three of which were useful. The trials for each group were carried out from April to September 2015 (autumn/winter) and from December 2015 to May 2016 (summer/autumn).

The varietal groups were represented by three commercial hybrids: Pegasus (Santa Cruz), Dominador (Salada) and Pioneiro (Italiano).

Natural pollination took place through the process of self-pollination without any restrictions on the free visitation of pollinating agents.

Mechanical pollination was carried out daily between 9 and 10 a.m. using an electric toothbrush to vibrate the flowers.

The plants were transplanted into the field, tutored in the Vertical System with a zip tie and spaced 0.6 m apart and 1.0 m apart. Cultivation was carried out as necessary and according to specific recommendations for tomato plants under a fertigation system (ALVARENGA, 2013). The fertiliser recommendations for the crop were made according to the interpretation of the soil analysis based on the 5ª approximation (RIBEIRO et al., 1999).

Yield components were assessed in terms of the number of fruits per plant and productivity. Fruits were classified according to their calibre, number and mass of large (NFG and MFG), medium (NFM and MFM) and small (NFP and MFP) fruits according to the Ministry of Agriculture, Livestock and Supply (2002) standards, as well as the fruit's transverse half-diameter according to shape: oblong for Salada and round for Santa Cruz and Italian. Oblong fruits are small (50-65 mm), medium (65-80 mm), large (80-100 mm) and giant (>100 mm) in size, while round fruits are classified as small (40-50 mm), medium (50-60 mm) and large (>60 mm).

Quality attributes were assessed by determining the following variables: soluble solids (TSS) expressed in °Brix, titratable acidity (TA) expressed in % citric acid, flavour, pH, firmness (FR) and lycopene content (L).

The soluble solids content (TSS), expressed in °Brix, was measured in the fruit pulp using a HANNA HI 96801 refractometer. Total titratable acidity (TA) was determined according to the methodology of the Adolfo Lutz Institute and expressed as a % of citric acid. Flavour was estimated using the ratio between total soluble solids and total titratable acidity. The pH was measured using a HANNA pH 20 phantom. Firmness (FR) was assessed using an Instrutherm DD-200 penetrometer at 2 locations in the equatorial region of the fruit and expressed in Newton. Lycopene (L) content was determined using a BEL PHOTONICS, SP1105 spectrophotometer at a wavelength of 470 nm.

The number of viable seeds per fruit (NSV) and the dry mass of 100 seeds (MSS) were obtained according to MAPA (2009) seed analysis rules, using a TE-391 BOD to analyse the number of viable seeds from the germination test and a TE394-2 oven to determine the dry mass.

The data relating to seed analyses, fruit quality and yield were analysed separately for each planting season using SISVAR software (5.6).

4.3.   RESULTS AND DISCUSSION

Production components in the Santa Cruz, Salada and Italiano groups.

Table 1 shows the mean square values and coefficients of variation for the variables related to fruit production for the Santa Cruz group. There was a significant effect in relation to the planting times considered when evaluating NFM, NFP, MFG, MFM, MFP and productivity.

Table 1. Summary of the analysis of variance for the variables number of large, medium and small fruits per plant (NFG, NFM, NFP); production of large, medium and small fruits per plant (MFG, MFM, MFP); total number of fruits per plant (NFT) and yield measured in the Santa Cruz tomato hybrid in the autumn/winter and summer/autumn planting seasons.

FV	GL	Mean squares							
		NFG	NFM	NFP	NTF	MFG	MFM	MFP	Productivity
Block	3	14.82ns	4.58ns	3.58ns	18.27ns	1209.50ns	221.27ns	1898,66	78.70ns
Season	1	6.59ns	61,66*	80,95*	1.98ns	18846,48**	11539,59**	7695,23*	992,09*
Pollination	1	0.46ns	2.15ns	2.64ns	14.3ns	243.28ns	373.55ns	1343.40ns	8.68ns
Season*Pollination	1	51.58ns	6.08ns	0.39ns	105.72ns	68.76ns	81.04ns	672.75ns	774,64*
Waste	9	16,23	11,98	4,99	41,68	1774,08	169,64	926,81	148.84ns
CV		28,24	46,39	49,64	24,62	23,49	10,32	43,25	19,49

ns Not significant at 5% probability by the F test; and ** significant at 1% probability by the F test.

The autumn/winter planting showed an increase in NFM (Figure 1). As for the variables NFP (Figure 2), MFG (Figure 3), MFM (Figure 4) and MFP (Figure 5), the average values for NFM were higher when the crop was grown in the summer/autumn planting season.

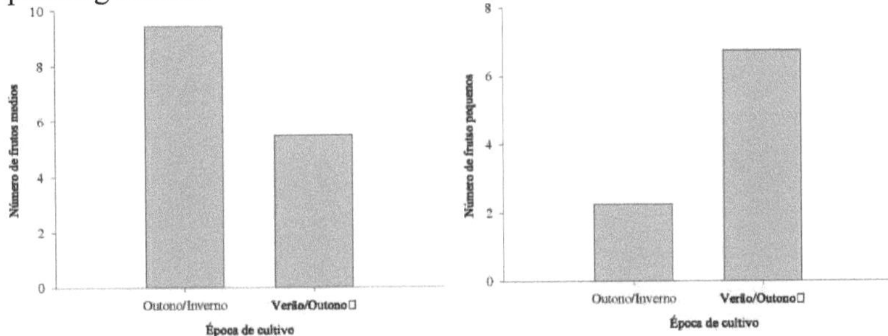

Figure 1. Number of average fruits (NFM) evaluated in the Santa Cruz (SC) group in the autumn/winter and summer/autumn planting seasons according to analysis of variance.

Figure 2. Number of small fruits (NFP) evaluated in the Santa Cruz group (SC) autumn/winter and summer/autumn planting seasons according to analysis of variance.

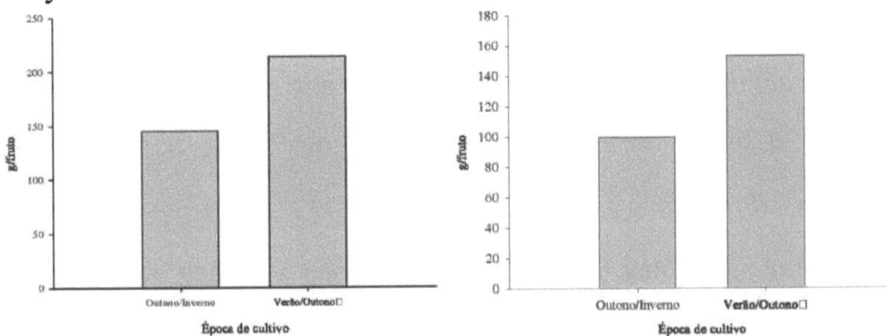

Figure 3: Mass of large fruit (MFG) evaluated in the Santa Cruz (SC) group in the autumn/winter and summer/autumn planting seasons according to analysis of variance.

Figure 4 - Average fruit mass (MFM) evaluated in the Santa Cruz (SC) group in the autumn/winter and summer/autumn planting seasons according to analysis of variance.

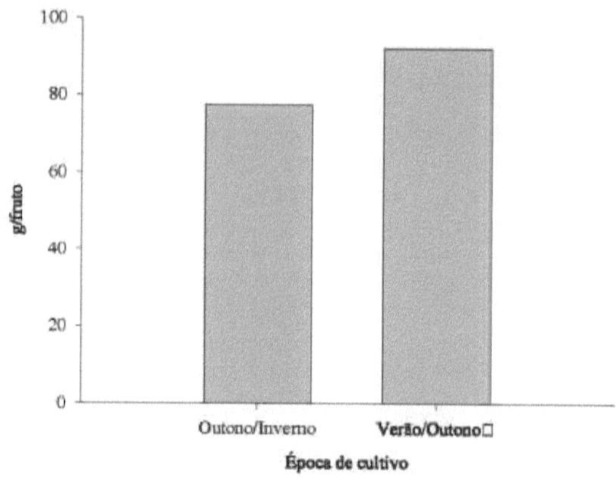

Figura 5. Mass of small fruits (MFP) evaluated in the Santa Cruz (SC) group in the autumn/winter and summer/autumn planting seasons according to analysis of variance.

Table 2 shows the averages for the planting times x pollination interaction, which suggests that the variation in production for the Santa Cruz group depends on the planting time and the type of pollination used. Mechanical pollination only had a positive effect in the autumn/winter season. These results also suggest that production in summer/autumn is higher than in autumn/winter, regardless of the type of pollination.

The increases in yield may be related to the greater vibrancy of the tomato blossom, a consequence of the effect of mechanical pollination in the autumn/winter and the greater activity of pollinating agents in the summer/autumn that lead to an increase in the mass of large (Figure 3), medium (Figure 4) and small (Figure 5) fruit at this time of planting.

Table 2. Production evaluated in the Santa Cruz (SC) group in the planting time x pollination interaction.

Planting time	Pollination	
	Natural	**Mechanics**
Autumn/Winter	47,44	56,44
Summer/Autumn	76,94	64,08

Low temperatures and high relative humidity in autumn/winter hinder the activities of pollinating insects and consequently the processes of pollination and fertilisation of the oosphere (**DAŞGAN** et al., 2004; PRESSMAN et al., 1999); in this sense, mechanical pollination as a complementary practice reduces these types of difficulties by increasing the number of pollen deposited on the flower's stigma. HARDER; BARCLAY, (1994) reported that mechanical pollination manages to release 14% more pollen in a single visit than that released by pollinating bees.

Table 3 shows the mean square values and coefficients of variation for the characteristics related to fruit production, evaluated in the Salada group. There was a significant effect of planting time for the NFP variable and pollination for the yield variable.

Table 3. Summary of the analysis of variance for the variables number of large, medium and small fruits per plant (NFG, NFM, NFP); production of large, medium and small fruits per plant (MFG, MFM, MFP); total number of fruits per plant (NFT) and yield measured in the Salada tomato hybrid in the autumn/winter and summer/autumn planting seasons.

FV	GL	Mean squares							
		NFG	NFM	NFP	NTF	MFG	MFM	MFP	Productivity
Block	3	32.92ns	5.87ns	16.8500ns	24.00ns	1071.20ns	1913.12ns	425.04ns	342.72ns
Season	1	14.1ns	0.0637ns	97,46*	34.45ns	34.45ns	1495.94ns	2150.17ns	259.37ns
Pollination	1	0.25ns	3.0537ns	65.00ns	86.67ns	86.67ns	4758.58ns	1046.84ns	864,06*
Season*Pollination	1	0.24ns	3.0712ns	43.06ns	28.19ns	28.19ns	4453.22ns	1512.43ns	344.47ns
Waste	9	15,69	5,34	17,94	19,39	19,39	1059,95	714,43	147,49
CV		57,10	34,55	38,84	17,95	15,58	19,98	27,58	20,24

ns Not significant at 5% probability by the F test; and ** significant at 1% probability by the F test.

The same effect was seen for the NFP variable (Figure 6) as for the Santa Cruz group (Figure 2), with a higher number of small fruits in the summer/autumn planting season. This increase may be due to the higher fruit set in the upper third. OLIVEIRA et al. (1995) states that smaller fruits are located after the first 4 bunches (upper third), and that these fruits are responsible for approximately 12.6% of the plant's total production.

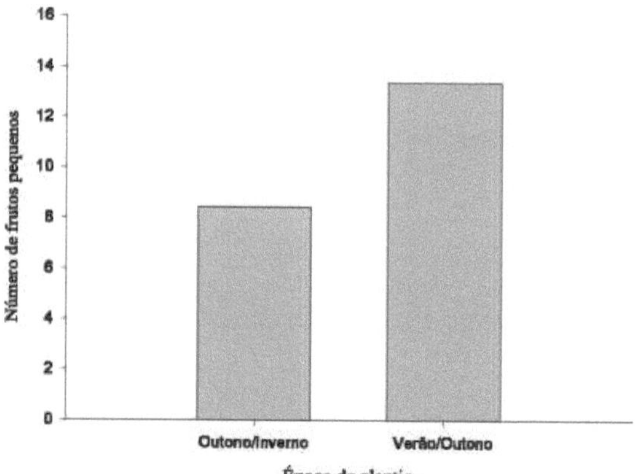

Figura 6. Number of small fruits (NFP) evaluated in the Salad group (S) in the autumn/winter and summer/autumn planting seasons according to analysis of variance.

Mechanical pollination in the Salada group increased yields by 22% (Figure 7). The Dominador tomato used in this trial, as a hybrid of the Salada group, has more dense leaves (OTONI et al., 2015) when compared to the other hybrids. This characteristic could create barriers that make it difficult for pollinating agents to access the flowers of the plants; therefore, regardless of the planting time, mechanical pollination increases productivity.

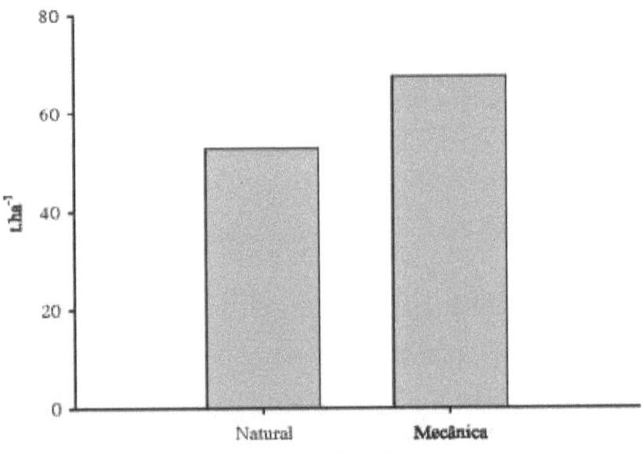

Figura 7. Productivity (t.ha^{-1}) evaluated in the Salada group (S) with natural and mechanical pollination according to analysis of variance.

Table 4 shows the mean square values and coefficients of variation for the characteristics related to fruit production, evaluated in the Italian group. All the variables except NFG and NFM showed a significant effect in relation to the time of planting, while for NFT, MFG and productivity significance was found for the type of pollination. There was a significant interaction for the NFP and MFP variables.

Table 4 Summary of the analysis of variance for the variables number of large, medium and small fruits per plant (NFG, NFM, NFP); production of large, medium and small fruits per plant (MFG, MFM, MFP); total number of fruits per plant (NFT) and yield (t ha^{-1}) measured in the Italian tomato hybrid in the autumn/winter and summer/autumn planting seasons.

FV	GL	Mean squares							
		NFG	NFM	NFP	NTF	MFG	MFM	MFP	Productivity
Block	3	63.73ns	1.73ns	9.06ns	28.29ns	617.97ns	189.87ns	308.99ns	265.62ns
Season	1	8.83ns	49.80ns	181,30**	308,17*	7439,92**	3275,27**	3927,84*	2552,77**
Pollination	1	17.28ns	96.28ns	19.44ns	337,64*	982,82*	37.82ns	312.31ns	1495,36*
Season*Pollination	1	9.96ns	17.53ns	37,08*	50.76ns	127.23ns	257.76ns	2380,22*	0.97ns
Waste	9	18,1	28,0	6,50	41,3	175,0	85,10	381,3	145,70

	0	0		0	0		0	
CV	30,92	50,46	44,16	29,98	7,83	7,85	30,78	18,44

ns Not significant at 5% probability by the F test; and ** significant at 1% probability by the F test.

The NTF (Figure 8), MFG (Figure 9), MFM (Figure 10) and yield (Figure 11) variables were higher in the summer/autumn planting season. The increase in productivity can be considered a consequence of the increase in fruit mass and number, as studies show that the total number of fruits produced (Figures 8 and 12) is related to the fruit set index on the plant (VIVIAN et al., 2015).

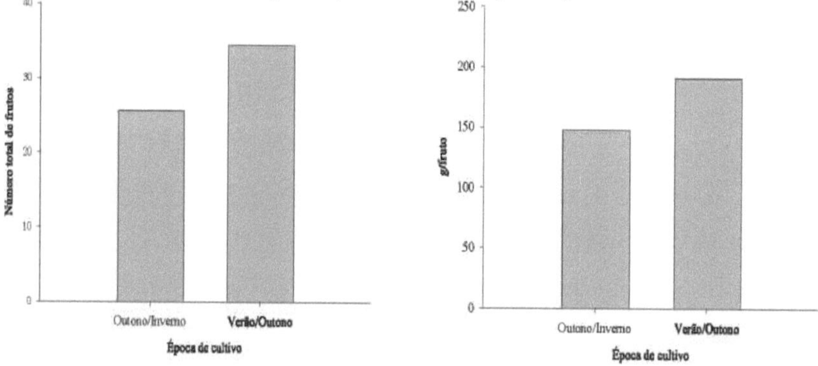

Figure 8. Total number of fruits (NTF) evaluated in the Italian group (I) in the autumn/winter and summer/autumn planting seasons according to analysis of variance.

Figure 9. Mass of large fruit (MFG) evaluated in the Italian group (I) in the autumn/winter and summer/autumn planting seasons according to analysis of variance.

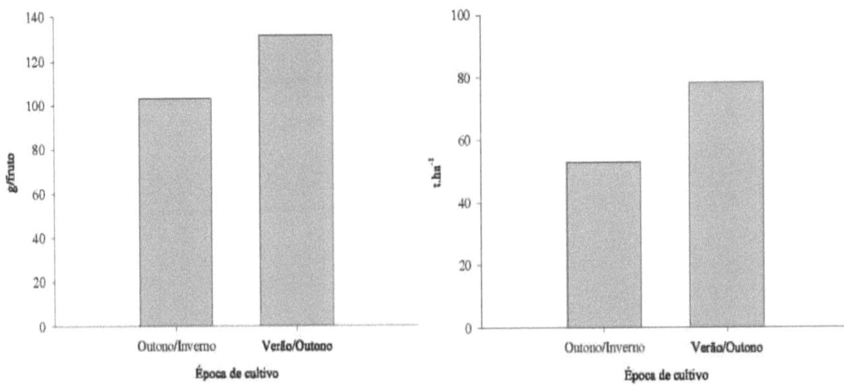

Figure 10. Average fruit mass (MF) evaluated in the Italian group (I) in the autumn/winter and summer/autumn planting seasons according to analysis of variance.
Figure 11. Productivity (t.ha^{-1}) evaluated in the Italian group (I) in the autumn/winter and summer/autumn planting seasons according to analysis of variance.

Mechanical pollination in the Italian group increases productivity (Figure 14) by 26 % when compared to natural pollination, regardless of the planting season. NTF (Figure 12) and MFG (Figure 13) are variables that increase productivity with the use of mechanical pollination (Figure 14).

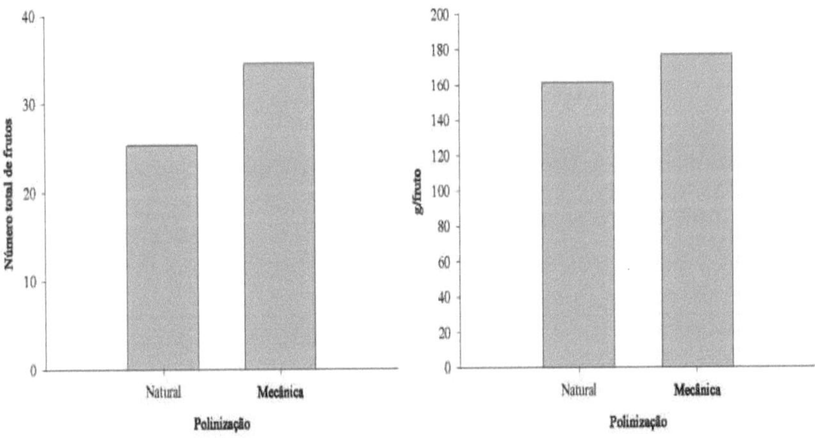

Figure 12. Total number of fruits (NTF) evaluated in the Italian group (I) in natural and mechanical pollination according to analysis of variance.

Figure 13. Mass of large fruit (MFG) evaluated in the Italian group (I) in natural and mechanical pollination according to analysis of variance.

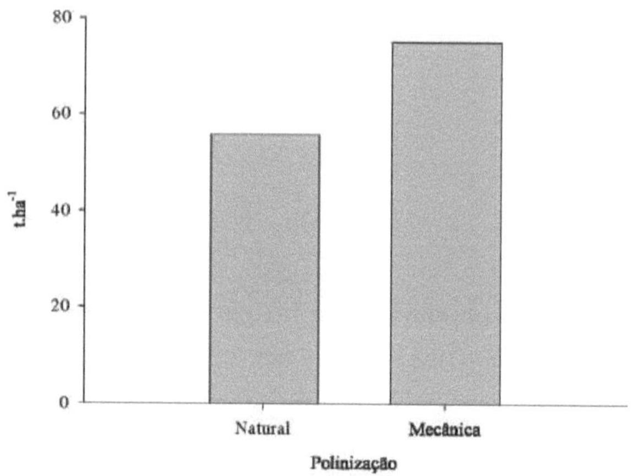

Figura 14.   Productivity (t.ha^{-1}) evaluated in the Italian group (I) with natural and mechanical pollination according to analysis of variance.

Table 5 shows the means of the season x pollination interaction for the NFP and MFP variables, which are dependent on the planting season together with the type of pollination. Mechanical pollination only positively altered the number of small fruits in summer/autumn, and had little influence on the increase in their number.

mass during the two planting seasons. The analyses also suggested that the mass and number of small fruits were higher during the summer/autumn planting.

Table 5. Number and mass of small fruits (NFP and MFP) evaluated in the Santa Cruz (SC) group in the planting time x pollination interaction.

Planting time	Pollination			
	Natural		Mechanics	
	NFP	MFP	NFP	MFP
Autumn/Winter	2,25	62,29	1,69	62,50
Summer/Autumn	6,00	90,00	11,50	71,65

According to VIVIAN et al. (2015), the number of fruits produced is directly related to the rate of fruit set on the plant. However, the higher yields for the Santa

30

Cruz and Italiano groups obtained during summer/autumn planting (Table 2 and Figure 11) may be the result of the increase in average fruit mass, as well as the increase in the number of fruits classified as small for the Santa Cruz group and the total number of fruits in the Italiano group.

Mechanical pollination and the summer/autumn planting season favour productivity, since mechanical pollination facilitates the release of pollen from the anthers, and in the summer/autumn season there is greater activity by pollinating insects as a result of the favourable environmental conditions in terms of temperature (23 °C) and average relative humidity (78 %) during flowering.

HIGUTI et al. (2010), studying the efficiency of pollination, also found higher fruit production from mechanical pollination, as a result of greater fruit set. For their part, CUÉLLAR et al. (2011), using electric vibration on tomato flowers, increased yield/plant by 34 %. These results were similar to ours, where yields increased by 22 % and 26 % for the Salada and Italian groups respectively, regardless of the season, and by 16 % for the Santa Cruz group in the autumn/winter season.

Quality attributes in the Santa Cruz, Salada and Italiano varietal groups.

The summary of the analysis of variance for the fruit quality variables for the Santa Cruz group is shown in Table 6.

Table 6. Summary of the analysis of variance for the variables total soluble solids (TSS), titratable acidity (TA), flavour, pH, firmness (FR) and lycopene (L) measured in Santa Cruz tomato hybrids during the autumn/winter and summer/autumn planting seasons.

FV	GL	Mean squares					
		SST	AT	Flavour	pH	L	FR
Block	3	0.06ns	0.00ns	1.42ns	0,02**	84.13ns	4.04ns
Season	1	0.25ns	0.04ns	11.42ns	0,06**	666.93ns	5.68ns
Pollination	1	0.00ns	0.00ns	0.33ns	0,0*	76.47ns	1.29ns
Season*Pollination	1	0.01ns	0.00ns	0.00ns	0.01ns	229.06ns	30.14ns

Waste	9	0,24	0,00	6,19	0,00	224.25ns 7,37
CV		11,58	20,12	26,15	1,26	41,23 26,86

ns Not significant at 5% probability by the F test; and ** significant at 1% probability by the F test.

There was a significant effect for the block, planting time (Figure 15) and type of pollination (Figure 16) for the pH variable, which indicates that environmental conditions may have influenced the results. However, the average values for the different groups and pollination systems are within the ideal range (4.2-4.7) according to the tomato nutritional composition table (ALVARENGA, 2013).

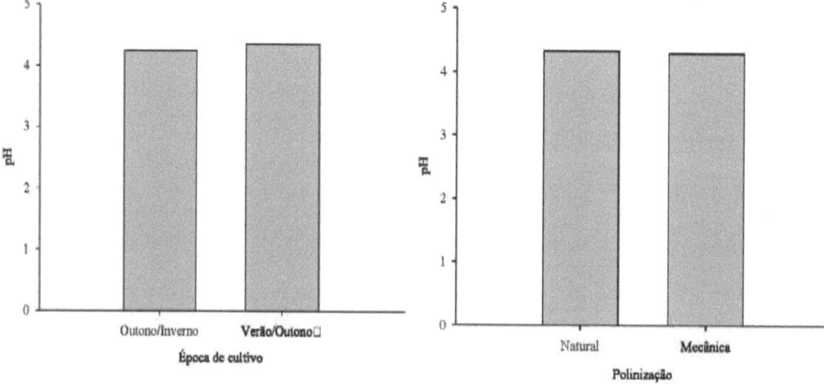

Figure 15. pH of fruit assessed in the Santa Cruz (SC) group in the autumn/winter and summer/autumn planting seasons according to analysis of variance.
Figure 16. pH of fruit evaluated in the Santa Cruz (SC) group in the natural and mechanical pollination system according to analysis of variance.

Table 7 summarises the analysis of variance for the fruit quality data in the Salada group. The time of planting influenced the variables AT, abor and lycopene content.

Table 7. Summary of the analysis of variance for the variables total soluble solids (TSS), titratable acidity (TA), flavour, pH, firmness (FR) and lycopene (L) measured in tomato hybrids of the Salada group in the autumn/winter and summer/autumn planting seasons.

FV	GL	Mean squares

		SST	AT	Flavour	pH	L	FR
Block	3	0.19ns	0.00ns	1.015ns	0.03ns	36.22ns	23.21ns
Season	1	0.03ns	0,03**	20,02**	0.03ns	868,42**	9.14ns
Pollination	1	0.03ns	0.00ns	0.59ns	0.00ns	296.76ns	27.24ns
Season*Pollination	1	0.00ns	0.00ns	1.23ns	0.01ns	215.86ns	16.60ns
Waste	9	0,05	0,00	2,40	0,01	104,98	15,22
CV		5,97	12,66	14,70	2,74	31,040	25,20

ns Not significant at 5% probability by the F test; and ** significant at 1% probability by the F test.

Figures 17, 18 and 19 show the average values for AT, flavour and lycopene content respectively at the different planting times. Despite the dissimilarity between the values, the ATT is in line with the recommendations for tomato plants. According to CANTWELL (2004), fruit should have values between 0.2 - 0.6 % citric acid.

For the flavour (Figure 18) and lycopene content (Figure 19) variables in the autumn/winter and summer/autumn seasons respectively, the values were lower than the averages established for tomatoes. According to MENCARELLI JR.; SALVEIT, (1988) tomatoes can be considered tasty when their flavour score is above 10. ALVARENGA (2012) argues that commercial tomato fruit should have an average of 30mg/kg of lycopene.

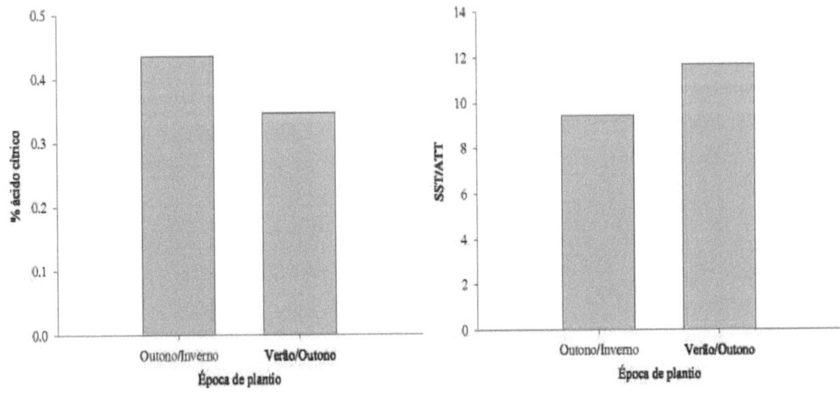

Figure 17. AT (% citric acid) of fruit assessed in the Salada group (S) in the autumn/winter and summer/autumn planting seasons according to analysis of variance.
Figure 18. Flavour (TSS/AT) of fruit from the Salad group (S) in the autumn/winter and summer/autumn planting seasons according to analysis of variance.

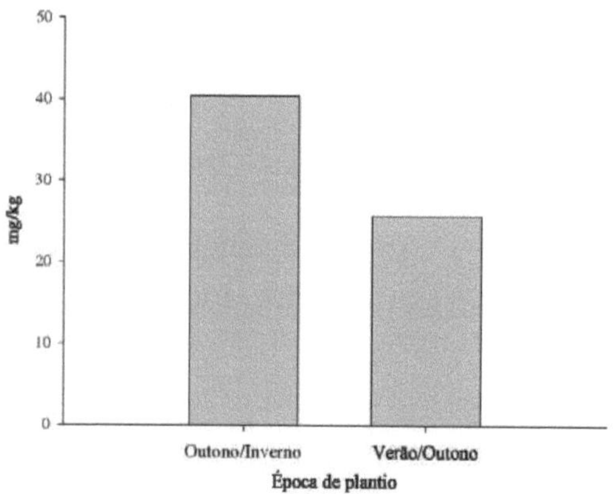

Figura 19. Lycopene content (mg/Kg) of fruit evaluated in the Salada group (S) in the autumn/winter and summer/autumn planting seasons according to analysis of variance.

Table 8 summarises the analysis of variance for the fruit quality data for the Italian group. Planting time influenced the variables TSS, TA, pH, L and FR.

Table 8. Summary of the analysis of variance for the variables total soluble solids (TSS), titratable acidity (TA), flavour, pH, firmness (FR) and lycopene (L) measured in Italian tomato hybrids in the autumn/winter and summer/autumn planting seasons.

FV	GL	Mean squares					
		SST	AT	Flavour	pH	L	FR
Block	3	0.23ns	0.00ns	2.61ns	0.01ns	134.97ns	22.68ns
Season	1	1,50*	0,03*	12.29ns	0,05*	608,34*	122,78*
Pollination	1	0.18ns	0.00ns	2.42ns	0.02ns	323.86ns	1.33ns
Season*Pollination	1	0.14ns	0.00ns	4.29ns	0.00ns	328.77ns	3.75ns
Waste	9	0,22	0,00	4,58	0,01	70,03	15,07
CV		10,05	14,71	15,75	2,46	30,86	38,35

ns Not significant at 5% probability by the F test; and ** significant at 1% probability by the F test.

Figures 20, 21, 22, 23 and 24 represent the behaviour of the variables with an effect on the planting season, where all of them, except pH (Figure 22), showed higher average values in autumn/winter. TSS (Figure 20), TA (Figure 21) and pH (Figure 22) are all within the recommended range for tomato plants, regardless of the season. CANTWELL (2004) states that TSS and TA should be between 3.5 - 7 °Brix and 0.2 - 0.6 % citric acid. ARANA et al. (2007) considers that the pH values for ripe tomatoes should be between 4 and 5. In turn, the lycopene content (Figure 24) and fruit firmness (Figure 23) in the summer/autumn season are below the average proposed for tomatoes, which is over 30mg/Kg for lycopene content (ALVARENGA, 2012), and 10 to 15 Newton (N) for fruit firmness (CANTWELL, 2004).

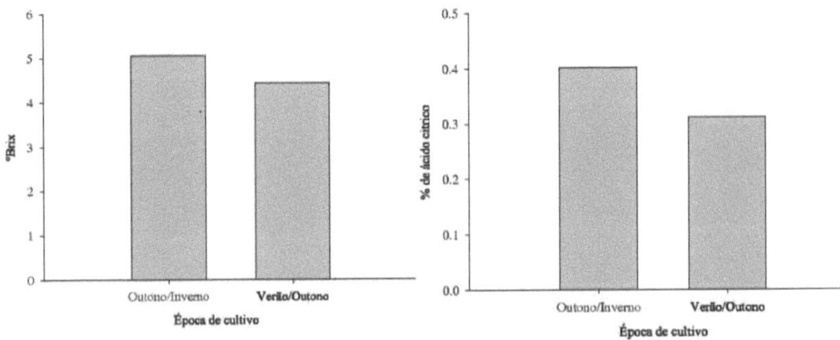

Figure 20. TSS (°Brix) of fruit evaluated in the Italian group (I) in the autumn/winter and summer/autumn planting seasons according to analysis of variance.
Figure 21. TA (% citric acid) of fruit assessed in the Italian group (I) in the autumn/winter and summer/autumn planting seasons according to analysis of variance.

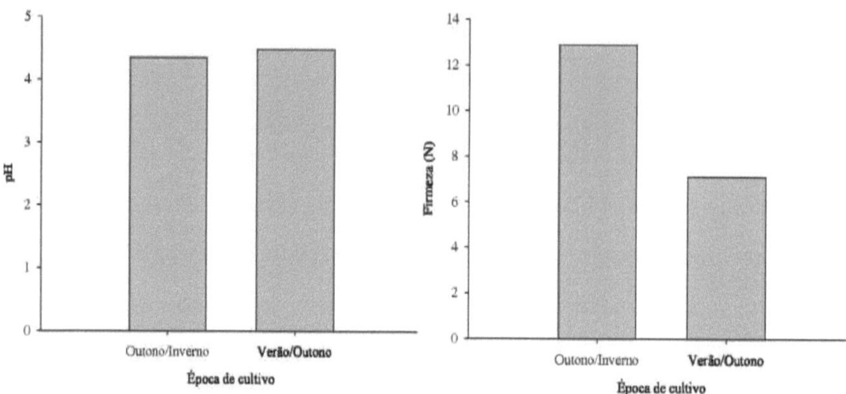

Figure 22. pH of fruit assessed in the Italian group (I) in the autumn/winter and summer/autumn planting seasons according to analysis of variance. Figure 23. Firmness (N) of fruit assessed in the Italian group (I) in the autumn/winter and summer/autumn planting seasons according to analysis of variance.

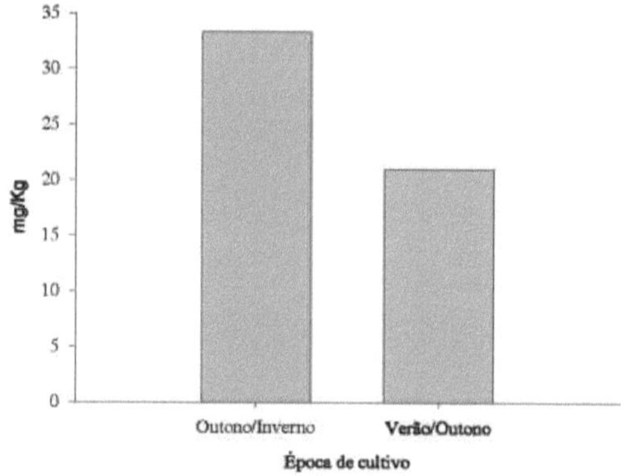

Figura 24. Lycopene content (mg/Kg) of fruit evaluated in the Italian group (I) in the autumn/winter and summer/autumn planting seasons according to analysis of variance.

In general, the type of pollination has no influence on tomato quality; only variables such as pH in the Santa Cruz and Italiano groups; TSS in the Italiano group; TA in the Salada and Italiano groups; flavour in the Salada group; lycopene content in the Salada and Italiano groups, and firmness in the Italiano group, are affected by the environment, with all except pH in the Santa Cruz and Italiano groups showing an increase during the autumn/winter planting season.

SILVA et al. (2010), argues that there are few studies that address quality in studies related to flower vibrations in tomato plants. However, some authors refer only to the soluble solids content in studies on buzz pollination processes with Bombus bees.

DOGTEROM et al. (1998) in Canada found no significant differences in soluble solids content when comparing treatments of non-pollinated plants and plants pollinated by bees. In similar studies carried out in Mexico, VERGARA; FONSECA-BUENDÍA (2012) found a significant effect between plants pollinated by bees and plants pollinated mechanically three times a week. This indicates that changes in fruit quality are dependent on environmental conditions and the frequency of days on which mechanical pollination practices are carried out. Further studies on fruit quality are needed in order to obtain consistent information to broaden our knowledge of the effects of mechanical pollination on fruit quality variables.

Evaluation of seeds in the Santa Cruz, Salada and Italiano varietal groups.

Table 9 summarises the analyses of variance for the seeds of the Santa Cruz, Salada and Italiano varietal groups. There was a significant effect for: the season for the NSV variable in the Santa Cruz group; the type of pollination and the interaction for the NSV variable in the Salada group; and the season in the Italiano group for the MSS variable.

Table 9. Summary of the analysis of variance for the variables number of viable seeds per fruit (NSV) and dry mass of 100 seeds (MSS) measured in tomato hybrids from the Santa Cruz, Salada and Italiano varietal groups in the autumn/winter and summer/autumn planting seasons.

FV	GL	Mean squares					
		Santa Cruz		Salad		Italian	
		NSV	MSS	NSV	MSS	NSV	MSS
Block	3	370.4ns	0.00ns	2409.83ns	0.01ns	137.97ns	0.00ns
Season	1	2090,0*	0.00ns	964.10ns	0.00ns	1064.06ns	0.03*
Pollination	1	1210.1ns	0.00ns	4579,22*	0.00ns	3492.21ns	0.00ns

Season*Pollination	1	568.9ns	0.00ns	4013,22*	0.00ns	87.89ns	0.00ns
Waste	9	258,2	0,00	746,44	0,00	1221.56	0,00
CV		10,2	22,42	20,25	37,85	26.66	22.42

ns Not significant at 5% probability by the F test; and ** significant at 1% probability by the F test.

Figure 25 shows the NSV variable, assessed during the autumn/winter and summer/autumn seasons in the Santa Cruz group. On average, a greater number of seeds were obtained for the autumn/winter planting.

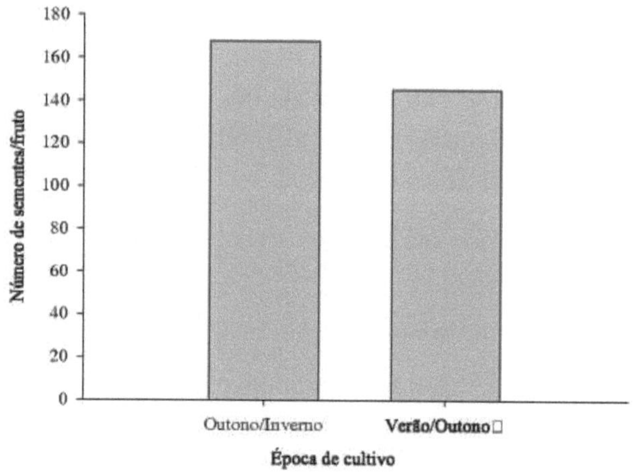

Figura 25. Number of viable seeds (NSV) per fruit evaluated in the Santa Cruz (SC) group in the autumn/winter and summer/autumn planting seasons according to analysis of variance.

In order to check the combined effect of the factors being evaluated, the interaction was split for the NFM variable in the Salada group in Table 10. Mechanical pollination increased the number of seeds in both planting seasons, but this type of pollination had a better effect in autumn/winter when compared to summer/autumn.

Table 10. Production evaluated in the Salada (S) group in the environment x pollination interaction.

Planting time	Pollination	
	Natural	**Mechanics**
Autumn/Winter	99,36	167,09
Summer/Autumn	137,25	145,13

The increase in the number of seeds is due to the low incidence of pollinating insects in autumn/winter. Mechanical pollination complements the natural pollination process. This practice favours the release of pollen that will be deposited on the stigma, which leads to an increase in the fertilisation rate of the oospheres and subsequent seed formation. Several authors have found that vibrating the flower causes a greater quantity of pollen grains to be deposited on the stigma, resulting in a greater number of seeds, size, weight and uniformity of the fruit (ALDANA et al., 2007; CUÉLLAR et al., 2011; SANTOS, 2014).

The increase in NSV in the Santa Cruz group in autumn/winter (Figure 25) and the use of mechanical pollination in the Salada group (Table 10) could also be related to the increase in productivity (Table 2 and Figure 7), where the greater the number of seeds developed, the greater the hormonal effect and consequently the fruit production. The literature reports a positive relationship between the number of seeds and productivity (CUÉLLAR et al., 2011; HIGUTI et al., 2010a; VERGARA; FONSECA-BUENDÍA, 2012)(CUÉLLAR et al., 2011; HIGUTI et al., 2010a; VERGARA; FONSECA-BUENDÍA, 2012)(DOGTEROM et al, 1998; HIGUTI et al., 2010; CUÉLLAR et al., 2011; VERGARA; FONSECA-BUENDÍA, 2012); however, in our study this relationship was not verified for the Italian group, where the number of seeds had no effect for the season or the type of pollination (Table 9).

Fruit development is regulated by the interaction of growth regulators that coordinate the different stages of tomato development, and the number of seeds influences their production (DORCEY et al., 2009; MEDINA et al., 2013; OBROUCHEVA, 2014; SERRANI; RUIZ-RIVERO, 2008). SHINOZAKI et al. (2015) argue that the process of pollination and subsequent fertilisation induces the biosynthesis of hormones that stimulate fruit formation.

Some studies have established that there is a relationship between hormonal activity and fruit quality (SAGAR et al., 2013; SU et al., 2015), while others indicate that this activity is related to the number of seeds (SERRANI; RUIZ- RIVERO, 2008). However, in our study there were no changes in fruit quality with an increase in the number of seeds as a result of the effect of mechanical pollination.

MSS was higher in the autumn/winter season when compared to the summer/autumn season in the Italian group (Figure 26). These results suggest that during the autumn/winter season, the seeds had a greater accumulation of substances such as proteins, sugars and lipids (DIAS, 2001).

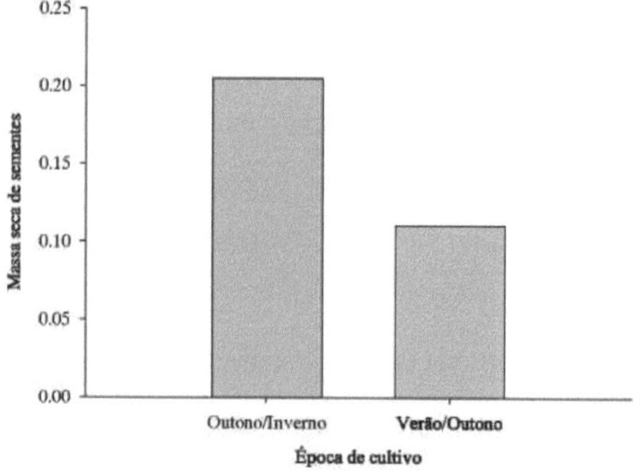

Figura 26. Dry mass of seeds (MSS) per fruit evaluated in the Italian group (I) in the autumn/winter and summer/autumn planting seasons according to analysis of variance.

It should be emphasised that mechanical pollination can be a sustainable practice that fits in with the majority of the groups evaluated, as it creates short-term economic benefits with minimal environmental effects, since its establishment does not require inverting considerable amounts of money or altering the usual cultural practices of tomato production systems.

4.4   CONCLUSIONS

Mechanical pollination as a complementary practice in crops grown in full sun is effective in the Salada and Italiano varietal groups, increasing yields by 22% and 26% respectively, regardless of the planting season, and in the Santa Cruz group during the autumn/winter season it increases yields by 16%.

There is no change in fruit quality with the use of mechanical pollination for crops grown in open fields.

4.5   **BIBLIOGRAPHY**

ALDANA, J.; CURE, J. R.; ALMANZA, M. T.; VECIL, D.; RODRÍGUEZ, D. Effect of the

de *Bombus atratus* (Hymenoptera: Apidae) sobre la productividad de tomate *(Lycopersicon esculentum* Mill.) bajo invernadero en la Sabana de Bogotá, Colombia. **Agronomia colombiana,** v. 25, n. 1, p. 62-72, 2007.

ALVARENGA, M. A. R. Tomato: production in the field, greenhouse and hydroponics. rev. 2.ed. **Lavras: Editora Universitária de Lavras,** 2013.

ARANA, I.; JARÉN, C.; ARAZURI, S.; GARCÍA, G. M. J.; URSUA, A.; RIGA, P. 2007. **Quality of fresh tomatoes**: cultivation technique and variety. Available at: <http://www.horticom.com/pd/imágenes/67/359/67359.pdf. > Accessed on 19/08/2016.

ARIIZUMI, T.; SHINOZAKI, Y.; EZURA, H. Genes that influence yield in tomato. **Breeding Science,** v. 63, n. 1, p. 3-13, 2013.

CANTWELL, M. Fresh market Tomato Statewide Uniform Variety Trial Report Field and Postharvest Evaluations. **South San Joaquin Valley.** UCCE. University of California Cooperative Extension. Michelle Le Strange, UCCE Farm Advisor, Tulare & Kings Counties, p. 25, 2004.

CARDOSO, A. I. I. Chilli production with plant vibration. **Ciência e Agrotecnologia,** v. 31, n. 4, p. 1061-1066, 2007.

CARRIZO GARCIA, C.; MATESEVACH, M.; E BARBOZA, G. Features related to anther opening in *Solanum* species (Solanaceae). **Botanical Journal of the Linnean Society,**

v. 158, n. 2, p. 344-354, 2008.

CUÉLLAR, J; COOMAN, A; ARJONA, H. Increasing the productivity of winter tomato cultivation by improving pollination. **Agronomía Colombiana**, v. 18, n. 1-3, p. 39-45, 2001.

DAÇGAN, H. Y.; ÔZDOGAN, A. O.; KAFTANOGLU, O.; Abak, K. Effectiveness of Bumblebee Pollination in Anti-Frost Heated Tomato Greenhouses in the Mediterranean Basin*. **Turkish Journal of Agriculture and Forestry**, v. 28, n. 2, p. 73-82, 2004.

DE, L. P. A.; VALLEJO-MARIN, M. What's the 'buzz'about? The ecology and evolutionary significance of buzz-pollination. **Current Opinion in Plant Biology**, v. 16, n. 4, p. 429-435, 2013.

DIAS, D. C. F. Seed ripening. **Federal University of Viçosa**. Available at: <http://www.seednews.inf.br/espanhol/seed56/artigocapa56_esp.shtml>.   Accessed on: 22 August 2016.

DOGTEROM, M. H.; MATTEONI, J. A.; PLOWRIGHT, R. C. Pollination of Greenhouse Tomatoes by the North American Bombus vosnesenskii (Hymenoptera: Apidae). **Journal of Economic Entomology**, v. 91, n. 1, p. 71-75, 1998.

DORCEY, E.; URBEZ, C.; BLÁZQUEZ, M. A.; CARBONELL, J.; PEREZ-AMADOR, M. A. Fertilisation-dependent auxin response in ovules triggers fruit development through the modulation of gibberellin metabolism in Arabidopsis. **The Plant Journal**, v. 58, n. 2, p. 318-332, 2009.

GILLASPY, G.; BEN-DAVID, H.; GRUISSEM, W. Fruits: a developmental perspective. **The Plant Cell**, v. 5, n. 10, p. 1439, 1993.

GOULSON, D. Bumblebees: behaviour, ecology and conservation 2 ed. **Oxford University Press**, 2010.

HARDER, L. D.; BARCLAY, R. M. R. The functional significance of poricidal anthers and buzz pollination: controlled pollen removal from Dodecatheon. **Functional Ecology**, p. 509-517, 1994.

HIGUTI, A. R. O.; GODOY, A. R.; SALATA, A. D. C.; CARDOSO, A. I. I. Tomato production in function of plant" vibration". **Bragantia**, v. 69, n. 1, p. 87-92, 2010.

MEDINA, M.; ROQUE, E.; PINEDA, B.; CANAS, L.; RODRIGUEZ-CONCEPCIÓN, M.; BELTRÁN, J. P.; GÓMEZ-MENA, C. Early anther ablation triggers parthenocarpic fruit development in tomato. **Plant Biotechnology Journal**, v. 11, n. 6, p. 770-779, 2013.

MELO, PAULO CT. Recent advances in table tomato growing associated with changes in the technological paradigm and challenges to overcome. **5th National Table Tomato Seminar (5th SNTM).** Available at <http://www.tomatedemesa.com.br/2014/noticia01.html>. UNIMEP, Piracicaba-SP, 2014.

MENCARELLI, F.; SALVEIT, J. R. Ripening of mature-green tomato fruit slices. **Journal of American Society Horticultural Science**, v. 113, p. 742-745, 1988.

MINISTRY OF AGRICULTURE, LIVESTOCK AND SUPPLY - MAP. Standards for the identification, quality, packaging and presentation of tomatoes. **Federal Official Gazette**, Brasilia, 2002.

NAHIR, D.; GAN-MOR, S.; RYLSKI, I.; FRANKEL, H. Pollination of Tomato Flowers by a Pulsating Air Jet. **Transactions of the ASAE**, v. 27, n. 3, p. 894-896, 1984.

OBROUCHEVA, N. V. Hormonal regulation during plant fruit development. **Russian Journal of Developmental Biology**, v. 45, n. 1, p. 11-21, 2014.

OTONI, B. D. S.; DA MOTA, W. F.; BELFORT, G. R.; SILVA, A. R. S.; VIEIRA, J. C. B.; DE SOUZA ROCHA, L. Production of tomato hybrids grown under different shading percentages. **Ceres**, v. 59, n. 6, 2015.

PEREIRA RAMOS, A. R.; ESTEVES AMARO, A. C.; MACEDO, A. C.; ASSIS SUGAWARA, G. S. D.; EVANGELISTA, R. M.; RODRIGUES, J. D.; ONO, E. O. Quality of 'giuliana' tomato fruit treated with products with physiological effects. **Semina: Ciências Agrárias**, p. 3543-3552, 2013.

PRESSMAN, E.; SHAKED, R.; ROSENFELD, K.; HEFETZ, A. A comparative study

of the efficiency of bumble bees and an electric bee in pollinating unheated greenhouse tomatoes. **The Journal of Horticultural Science and Biotechnology**, v. 74, n. 1, p. 101-104, 1999.

RIBEIRO, A. C.; GUIMARÃES, P. T. G.; ALVAREZ, V. H. (Ed.). **Recommendations for the use of correctives and fertilisers in Minas Gerais - 5ª** approximation. Viçosa, MG: CFSEMG, 1999.

SAGAR, M.; CHERVIN, C.; MILA, I.; HAO, Y.; ROUSTAN, J. P.; BENICHOU, M.;

PECH, J. C. SlARF4, an Auxin Response Factor Involved in the Control of Sugar Metabolism during Tomato Fruit Development. **Plant Physiology**, v. 161, n. 3, p. 1362-1374, 1. 2013

SANTOS, A. O. R.; BARTELLI, B. F.; NOGUEIRA-FERREIRA, F. H. Potential pollinators of tomato, Lycopersicon esculentum (Solanaceae), in open crops and the effect of a solitary bee in fruit set and quality. **Journal of Economic Entomology**, v. 107, n. 3, p. 987-994, 2014.

SERRANI, J. C.; RUIZ-RIVERO, O.; FOS, M.; GARCÍA-MARTÍNEZ, J. L. Auxin-induced fruit-set in tomato is mediated in part by gibberellins. **The Plant Journal**, v. 56, n. 6, p. 922-934, 2008.

SHINOZAKI, Y.; HAO, S.; KOJIMA, M.; SAKAKIBARA, H.; OZEKI-IIDA, Y.; ZHENG, Y.; OKABE, Y. Ethylene suppresses tomato (Solanum lycopersicum) fruit set through modification of gibberellin metabolism. **The Plant Journal**, v. 83, n. 2, p. 237-251, 2015.

SILVA, P. N.; HRNCIR, M.; FONSECA, V. L. I. Pollination by vibration. **Oecologia Australis**, v. 14, n. 1, p. 140-151, 2010.

SRIVASTAVA, A.; HANDA, A. K. Hormonal regulation of tomato fruit development: a molecular perspective. **Journal of Plant Growth Regulation**, v. 24, n. 2, p. 67-82, 2005.

SU, L.; DIRETTO, G.; PURGATTO, E.; DANOUN, S.; ZOUINE, M.; LI, Z.; CHERVIN, C. Carotenoid accumulation during tomato fruit ripening is modulated by the auxin-ethylene balance. **BMC plant biology**, v. 15, n. 1, p. 1, 2015.

TOHGE, T.; FERNIE, A. R. Metabolomics-Inspired Insight into Developmental,

Environmental and Genetic Aspects of Tomato Fruit Chemical Composition and Quality. **Plant & cell physiology**, v. 56, n. 9, p. 1681-96, sep. 2015.

VANBERGEN, A. J. Threats to an ecosystem service: pressures on pollinators. **Frontiers in Ecology and the Environment**, v. 11, p. 251-259, 2013.

VELTHUIS, H. H. The historical background of the domestication of the bumble-bee, Bombus terrestris, and its introduction in agriculture. **Pollinating Bees-The conservation link between agriculture and nature. Ministry of Environment, São Paulo, Brazil**, p. 177-184, 2002.

VERGARA, C. H.; FONSECA-BUENDÍA, P. Pollination of greenhouse tomatoes by the Mexican bumblebee Bombus ephippiatus (Hymenoptera: Apidae). **Journal of Pollination Ecology**, v.7, p. 27-307, 2012.

VIVIAN, R.; ROCHA, A.; GALVÃO, H. L.; MARTINEZ, H. E. P.; PEREIRA, P. R. G.; FONTES, P. C. R. Planting density and number of leaves influencing yield and fruit quality of tomato plants grown with one cluster in a hydroponic system. **Ceres**, v. 55, n. 6, 584-589, 2015.

More
Books!

Printed by Books on Demand GmbH, Norderstedt / Germany